2014 - World on the Verge of Global Changes

Announcement of soon-to-be-published book
"Fundamentals of Predictable
Economy" by Yury G. Geltser

YURY G. GELTSER

authorHOUSE®

AuthorHouse™ LLC
1663 Liberty Drive
Bloomington, IN 47403
www.authorhouse.com
Phone: 1-800-839-8640

Published by AuthorHouse   05/16/2014

ISBN: 978-1-4969-1224-4 (sc)
ISBN: 978-1-4969-1225-1 (e)

About the Author

Yury G. Geltser was born in 1952 in Russia. He is a well-known businessman and scientist. Mr. Geltser is a candidate of Technical Sciences and a full member (academician) of the St. Petersburg Academy of Engineering in the department of Economics and Management in Engineering. Mr. Geltser is a full member of the Academy of Investment and Construction Economy. He is married and has two children. In his spare time Mr. Geltser enjoys reading as well as playing piano and sports.

Contents

2014—World on the Verge of Global Changes.

Today we would like to announce the book by Yuriy G. Geltser "Fundamentals of Predictable Economy" scheduled to be published in 2014. I have assumed a task to present the science of Economics through the language of General System Theory. Method of monitoring the system life cycles is far from the only methodological principle of this theory. However, it is the very method I would like to introduce to my readers by publishing the excerpts from the yet-to-be-published book. This is because, first and foremost, it touches upon the events close and dear to us, and, secondly, it brilliantly demonstrates the possibility of integration of two disciplines into one.

In lieu of preamble

IN ESSENCE, WHAT IS THE CASE?

"The decisive question is that of the explanatory and predictive value of the "new theories" attacking the host of problem around wholeness, teleology, etc. Of course, the change in intellectual climate which allows one to see new problems which were overlooked previously, or to see problems in a new light, is in a way more important than any single and special application. The

"Copernican Revolution" was more than possibility somewhat better to calculate the movements of the planets; general relativity more than an explanation of a very small number of recalcitrant phenomena in physics; Darwinism more than a hypothetical answer to zoological problems; it was the changes in the general frame of reference that mattered (cf. Rapoport, 1959a). Nevertheless, the justification of such change ultimately is in specific achievements which would not have been obtained without the new theory."[1]

". . . [Th]e predictions would be reliable in as much as the conceptual representation of the world is truthful."[2]

"Dynamic systems studies usually are not designed to predict what will happen. Rather, they're designed to

---

[1]   Ludwig von Bertalanffy, General System Theory: Foundations, Development, Applications (New York: George Braziller, 1968), 99-100.

[2]   Masanao Toda, Emir H. Shuford, Jr., "Logic of Systems: Introduction to a Formal Theory of Structure," General Systems: Yearbook of the Society for the Advancement of General Systems Theory, vol. 10 (1965): 25.

explore what would happen, if a number of driving factors unfold in a range of different ways."[3]

"The future can't be predicted, but it can be envisioned and brought lovingly into being."[4]

In a world that surrounds a human the economy takes, perhaps, the most significant place. The world of production of goods, the world of machines and industrial relations. The world of conflict of material interests of individuals, groups and entire classes. The world of integration, globalization and international division of labor. The world of taxes, tariffs, protective duties, devaluations, inflation, crises and defaults.

This is a huge system with an incredibly complex structure. For at least the last three centuries, thousands of scientists have made attempts to study this system and learn how to manage it or at least predict its behavior.

This science has its own language, its own concepts. It operates a variety of the identified laws. This science attempts to formulate various

---

[3] Donella H. Meadows, Thinking in Systems: A Primer (White River Junction, VT: Chelsea Green Publishing, 2008), 46.

[4] Donella H. Meadows, quoted in ibid, 169.

forecasts, to regulate inflation and population employment, to stimulate the business world activity and protect it from "overheating."

However, economic crises are still a necessary evil, and they always come unexpectedly. Admittedly, many scientists may not agree to this. Well, as they may say, "We have long discovered the cycle of production and, if the economy is booming, then one may expect an imminent crisis followed by stagnation and a new upsurge thereafter." Everything is very simple. This is how it happens. Once the economy is delayed, even for a little while, at the point of growth, some economists begin predicting a forthcoming crisis. Is it any wonder that some of them become the prophets?

However, I have already pointed out once[5] that many scientists interpret the term "cyclicality" quite loosely. It is not each recurrence of the phenomena of nature or social life that may be called cyclical. First, recurrence should possess a relatively verified temporal character, such as, for instance, seasons. Deviations in one way or the other are possible as long as they are within the explainable limits. Second, to identify recurrence of the phenomena does not in itself mean to determine the cyclical processes until one will have identified the source causing such recurrence. For example, there lies the rotation of Earth around its axis

---

[5] Y. G. Geltser, "Cycles and Their Asymmetry: Economic Aspect," Journal Concepts, no. 2(21) (July 16, 2008): 98.

at the heart of alternation of day and night, and there lies the rotation of Earth around the Sun at the heart of the changing seasons.

In contrast, at the heart of economic cycles, as they are viewed by the science of Economics today, there lies nothing more but fixation of the economic ups and downs, and identification of the sufficiently broad spectrum of underlying causes of their origin.

For instance, what invaluable truths may be drawn from these findings? "The duration of long[6] cycles, from the initial to ultimate depression varies between the minimum of 6 years and the maximum of 13 years. If one calculates all small peaks and depressions along with the major cycles, then 18 cycles that occurred from 1865 to 1938 (if calculated this way) range from the minimum of 2 years to the maximum of 9 years in duration from depression to depression. The average duration is 3 years. 13 cycles out of 18 are within the range covering the period from three to five years."[7] Wherein, the cycle is intended to mean a temporal period of the "phase of upsurge and recession."[8]

---

[6] The scientific world of various countries does not offer consistency in use of terminology and, in this case, a medium-length, Juglar Cycle is being discussed.

[7] Hansen, quoted in L. E. Grinin, A. V. Korotayev, S. V. Tsirel, Development Cycles of Modern World System (Moscow, Russia: Publishing House Librokom, 2011), 13.

[8] L. E. Grinin, A. V. Korotayev, S. V. Tsirel, quoted in ibid, 13.

The authors have thoroughly questioned what the source of these cycles is. Here is their answer. ". . . The only real factor that can set the rhythm of fluctuations of this duration for the Kondratieff Waves[9] and their phases is the Juglar Cycles."[10] By the way, what is the source of the Juglar Cycles? The answer is as follows. "There is no clear explanation for the duration of the Juglar Cycles of 7-10 years."[11] Now the circle is complete.

Nevertheless, the study of economic history and all of human history in general from the perspective of recurrence of processes is unambiguously important. Ultimately, it is necessary to identify the patterns of such recurrences. A group of authors[12] are examining the history of Russia in the 400-year cycles, breaking them down into five 80-year cycles. However, here one may see the opposite extreme. Whatever the source of the human life cycles might be, these cycles cannot be rigidly fixed by certain numbers, especially within such significant time periods.

---

[9]   These are periods of 40-60 years.

[10]   L. E. Grinin, A. V. Korotayev, S. V. Tsirel, Development Cycles of Modern World System (Moscow, Russia: Publishing House Librokom, 2011), 123.

[11]   L. E. Grinin, A. V. Korotayev, S. V. Tsirel, quoted in ibid, 129.

[12]   See Kuzyk, et al., Russia in Space and Time (History of the Future) (Moscow, Russia: Institute of Economic Strategies, 2004).

While investigating the issue of economic time,[13] I make attempts to understand what the source of economic cycles is. In the meantime, let's note that the economy has two main intertwined cycles: the cycle of reproduction and the investment cycle. The latter may be divided into the innovation cycle and the actual investment cycle. Inability to manage them and control their interaction brings forth the crises.

Third, the cyclical recurrence of processes suggests comparability of the phenomena and consistent chain of interrelated events. Every economic crisis has its own causes of origin, depth of shocks, duration and history of overcoming. In this sense, recurrence of crises is highly relative and is based on fairly general criteria. Moreover, if one considers the causal relationships and the essence of the crisis phenomena, they differ as either the crises of overproduction, financial crises (defaults and devaluations), bank crises (chain of bank failures), stock market crises (the peak of the bubble) or as a result of speculation in land and real estate. Presently, the global economy is experiencing a debt crisis.

Fourth, I am absolutely not convinced the crises are indeed the significant milestone for justification of the cyclical economic processes. Science may truly consider itself science from the moment it is able to

---

[13]   See Chapter 3, Paragraph §6 B), infra.

predict with some certainty the most important events for itself. In this sense, the scientific nature of Economics has become a byword.[14]

Economists are compared with the generals who are always preparing for the war gone by. The only difference is that they are always preparing for a crisis gone by. As they do so, there appears the entire layer of scientists who consider this mission impossible, thus meaningless. This is, in a sense, true. It is difficult to look for a black cat in a dark room, especially if it is not there. After all, the question is what level of predictability is at issue. In the event you are waiting for forecasts in the dynamics of the stock market prices for the decades to come, I am not the right person for you. There are experts and specialists in certain types of commodities. It is your business to believe them or not. Personally, I'd be careful if I were you.

There exist many parameters, the fluctuation of which does not disturb stability of the system within certain limits. It is necessary to study these fluctuations. It is possible and necessary to predict them

---

[14] See S. M. Guriev, Myths of Economy. Misconceptions and Stereotypes Distributed by the Media and Politicians (Moscow, Russia: Alpina Business Books, 2006), 206; A. V. Ostalskiy, A Brief History of Money (St. Petersburg, Russia: Commercial Publishing House Amphora, 2010) 215, 217-218; A. B. Kobyakov, M. L. Khazin, *Sunset of the Dollar Empire and the End of "Pax Americana"* (Moscow, Russia: Veche, 2003), 440.

from the short-term perspective. This is also an important element of predictable economy.

However, what economic system may be called sustainable? What is sustainability? Could it be at all possible for a developing system to be sustainable? Is it possible to set a direction for this dynamic as well as for its parameters? Or, should we rely solely on the "hidden role of the market" and instead entrust the task of crises prediction to fortunetellers who make predictions on coffee grounds? Are all crises really crises? What should be considered as crises? What is the depth of our understanding of economic processes? How far in terms of time will this depth allow us to see into the future? In general, is it not the high time for us to build (organize) the economic relations (processes) so that we understand they are moving from point A to point B based on the specified and clear route (path)? Or, should we instead prove it is utopia and put an end to this?

Attempting to answer these questions is the goal of this study. How well I can do this is up to my readers to judge.

It may seem I have set the goal, which is mathematical in nature. It is possible that over time the mankind will learn how to solve such problems by using computer technology. If not completely, then at least partially. However, my goal is to ensure one approaches these problems conceptually based on causal relationships, logic and consistent scientific methodology.

Let me be forward, this is not a recipe book for forecasts nor a monologue of a "psychic." This is just an attempt to identify the ways to recreate predictable economy. Nothing more. At the same time, it is a question of a new paradigm in the science of Economics.

The essence of this paradigm is that the economy is treated not as a set of identified laws, but rather as an integrated system, all the structures and patterns of which must conform to this integrity and interact with one another. Moreover, the economic system is not treated by us simply as an open, self-sustaining system, but rather as a system that has reached such a level of development that it can no longer fail to be a goal-oriented system. Order of specific goal, their priorities, balanced resources, interception of unnecessary which prevents, or fails to contribute to, achieving the goals, sets forth a task of "building" the economic relations instead of "drifting on the surface." This, in turn, must make the economy transparent and clear in its development. This radically changes the notion of crisis in and of itself. They will begin to understand the term "crisis" as deviation from the set of goals or deceleration in their achievement in time rather than a decrease of certain individual indicators, which often do not reflect the objective reality. This book will discuss the necessity and possibility of such economy in the realm of dominance of private property.

. . . .

Chapter 3. Philosophical aspect.

. . . .

§6 Space and time.

. . . .

B) Economic time.

Starting from a certain historical moment, a thinking human begins to see the objects of being rather than just things. He or she begins to see the processes. These processes have a beginning and an end. A gap between them has a certain duration. Attempts to measure this duration lead to the concept of time. To arrive at this concept a human had to make another discovery—the cyclical recurrence of many processes, especially noticeable in cosmological observations.

Initially, a human operates empirically by relative time, which corresponds to certain rhythms and cumulative duration of various processes. However, there exist many processes and they are diverse. This makes one search for something that may be understood as the absolute time.

This dimension is being sought in the length of a day and night (rotation of Earth around its axis) and the duration of a year (rotation of Earth around the Sun). Newton's absolute time carries a slightly different notion. "Absolute time in Astronomy differs from ordinary solar time by the equation of time. For the natural solar day and night, ordinarily measured as equal, are actually unequal from one another. This inequality is in fact corrected by astronomers to apply more accurate time while measuring movement of the celestial bodies. Perhaps [in nature] there does not exist any such uniform process, which could be applied to measure time with perfect accuracy. All processes can be accelerated or slowed down; however, the run of absolute time may not be changed. Duration or length of existence of things is the same regardless of whether the movements are fast (which are applied to measure time) or slow, or even exist at all."[15]

Rejection of the Einstein's absolute time is based on the postulate that the speed of light is limited by certain boundaries and, most importantly, that there is a different approach to the definition of the simultaneity of events. According to the Einstein's definition, the simultaneity of events may be recorded only if these events are interrelated. Because, according to Einstein, there is no universal or privileged reference frame, based on which all events would interrelate, no system is better than the other (that is, not absolute, thus there is no absolute time). In my opinion, the latter claim is debatable because all measurements implemented by the

---

[15]   Isaac Newton, "Mathematical Principles of Natural Philosophy," quoted in Y. B. Molchanov, Four Concepts of Time in Philosophy and Physics (Moscow, Russia: Publishing House Science, 1977), 54.

mankind are the subject of arrangement. Absolute time is, of course, a relative concept; however, it is amenable to regulation.

Everything related to high speeds and space has the general cognitive character for the world of the science of Economics. To that end, I do not consider it appropriate to labor the point. In my opinion, V. I. Vernadsky has made a more meaningful contribution to the concept of the relativity of time. He and Bergen before him, spoke about certain real time as an imminent reflection of movement inherent in each type of matter. In 1929 Vernadsky introduced the concept of biological time.[16]

Later he wrote, "We are talking about the historical, geological, space, and other times. It is convenient to distinguish biological time, within which the phenomena of life develop."[17]

Biological time is different because of its internal substance. This is not the hours and minutes but rather the end of generational change. "The end defines the shortest time required to create a new organism, i.e. not only to create it (organization), but also to create all complex

---

[16] V. I. Vernadsky's works concerning the study of time started being published in 1975 only. Accordingly, the priority of use of the concept "biological time" is nowadays given to French histologist Lecomte du Noüy, who introduced this concept in 1936.

[17] V. I. Vernadsky, Problems of Biogeochemistry (Moscow, Russia: Publishing House Science, 1980), 273.

chemical bodies—proteins, etc., which must be created by the organism. Obviously, this is a natural phenomenon."[18]

Scientific discipline birthed by the concept of biological time became known as "biorhythmology." It studies a variety of reactions of different organisms to the external as well as their own internal rhythms.

G. P. Aksenov, who studied the works of V. I. Vernadsky, has written, "Vernadsky draws our attention to the works of J. Alexander who has two concepts: the concept of instant points and differentiation between the concepts of changes and movements. Vernadsky said, 'If in Newton times one described the movement, the new Physics focused upon internal changes of substance as well, wherein one series of movements is replaced by others.' 'This is why it is so important to follow the development of Philosophy,' as Vernadsky summed up this most important, original, fundamental part of his treatise.[19] It had managed to approach the concept of uniform space time before the science did. It set aside the concept of absolute time and space and raised the question regarding scientific study of the intrinsic properties of space time. Finally, "It gives the science a number of indications for collection of facts not blindly, but rather based on working, solid and deep hypothetical constructions (§ 49)."[20] As I can

---

[18] V. I. Vernadsky, quoted in ibid, 274.

[19] "About life (biological) time."

[20] G. P. Aksenov, V. I. Vernadsky, About Nature of Time and Space (Moscow, Russia: Publishing House Librokom, 2012), 148.

see, the Newton's absolute time is criticized from quite a different angle than the Einstein's.

The fact that the concept of irreversibility had replaced the observable, purely cyclical sustainable phenomena became another factor for rejecting absolute time. They started treating the following processes as irreversible: 1) radioactive, 2) evolution of stars, 3) history of the face of our planet—Earth's crust, 4) evolution of the species of animal matter, 5) generational change within the same species, and 6) historical process of the mankind.

So, what is the economic time like? I believe there is no uniform answer to this question today. Let me consider the options and discuss the possibility to hear the economic music of time in various rhythms.

As I have pointed out in the beginning of the book, the economic cycles occurring among the crises may not be used as the basis the way they are interpreted by either Juglar or Kuznetsov because, even if there is something objective therein, this is the rhythmic failures of various cycles. I am, however, interested in what actually lies at the heart of these cycles.

1. <u>The life cycle of means of production.</u>

This problem has recently become the subject of research. I devoted my book to this question through the example of aquatic structures.[21]

---

[21] See Y. G. Geltser, System and Technical Fundamentals of a Life Cycle of Engineering Design of Aquatic Structures (Moscow, Russia: SvR-ARGUS, 2006).

This problem became of interest (including myself) primarily from the viewpoint of technical and economic positions. How to extend, and whether it is worth extending, a life cycle of technical equipment and engineering structures and, most importantly, whether it is necessary to consider, at the design stage of constructing a project, how easy and cost-efficient it will subsequently be to dismantle and dispose of it. What will we leave behind for our offspring? Will this be the crumbling monuments of standing monsters and rotting dumpsters of toxic waste or a blooming planet? Another problem being solved in these studies is how to make the engineering structures capable of withstanding both the natural and the man-made disasters.

From the standpoint of the economy and its temporal rhythms, I am interested in something else. I am interested in the total life cycle of technical means defining one or another level of the economy. I need to know this by the industry sections and regions. I need to understand when and what will replace what we have today. Reproductive and investment cycles, financial and fiscal policies, science and education should be devoted to this.

One can approximate the timing of this cycle. Changes in the active technological means, such as machinery, equipment, etc. are implemented every 5-10 years. At any rate, these are the norms of amortization. Obsolescence sometimes accelerates this process. On average, every 20 years there are new generations of machinery radically different from the previous types. Buildings and structures are renovated every 50-100 years.

2. <u>The life cycle of revolutionary inventions</u>.

Karl Marx's popular expression that the economic epochs differ not by what is produced, but how it is produced, and what instruments of labor are used is rather exaggerated. In fact, the ships and locomotives, telegraph and an electric light bulb have defined one technical revolution, whereas the automobiles and airplanes, radio and television have defined another one; and space, computers, jet technology and new means of communication have identified the third one. If these three phases are merged into one, it will be the revolution of engines and communications. From a steam engine to a jet engine to the telegraph to the Internet there lies an interval of 130-150 years. That is, revolutionary innovations take place in these areas every 45-50 years. However, if one takes a closer look at, for example, the automotive or telecommunications technology, here one may also see radical changes every 20-25 years or sooner.

3. <u>The process of reproduction, including innovation and investment period</u>.

Here, I consider the period from the moment of "throwing" money into the new production until the moment of its full recoupment. So, taking into account the design, construction, research and development phases, it takes on average 7-10 years. However, the life of objects is not limited to their recoupment term. The life obeys the terms of radical technological change, which, as I have pointed out, does not occur more frequently than once in every 20-25 years.

4.   Generational changes.

In fact, it is the biological time in the economy. It is commonly assumed by demographers that generational changes occur every 20-25 years. This is a time period during which there occurs a change of the elites, substitution of certain professionals by others.

5.   Horizon of anticipation.

Horizon of prognostic anticipation is limited today to 20-50 years. Horizon of planned anticipation is defined by the Soviet school of planning purely empirically at 5-7 years, subject to annual adjustment. In Western economies, there is sometimes a limit at 2 years.

In my opinion, horizon of prognostic planning must be simply tied to the periods of generational change, engineering and technological change, that is, similarly 20-25 years.

Out of the aforementioned cycles, I envision the biological time of 20-25 years as having the decisive influence upon the economic and technical processes. This is the shortest limit, which, under definition of

V. I. Vernadsky, allows not only to change one generation by the other, but also organize a new economic space for the newcomers.

Understanding this rhythm may have very practical implications in setting the targets and forming the development programs. Looking into the future, a human predetermines his or her current behavior. "This is a key point because, in this case, a two-way intertemporal connection is created: not only the past determines the present but also the future impacts the present."[22]

Time is divided into work and leisure. This is understandable. Similarly, it is logical to divide it into the regular economic time and time of disaster prevention. If the former may be characterized as "time is money," the latter may be characterized as "time is more expensive than money." Prevention of accidents and natural disasters, preparation of the economy for military action, actions to prevent social unrest may serve as examples of the foregoing.

Time of the disaster prevention is frequently interconnected in the past with time of the foregone opportunities. As seen personally and, alas, not only by the author of these lines, the Russian economy has existed in this time for the past 10-12 years. Inactivity, lack of

---

[22]   E. V. Balatsky, "Evolution of the Concept of Time in the Science of Economics," Federal Online Edition "Capital of the Country" (May 18, 2010): 8-9. http://www.capital-rus/articles/article/177289

professionalism, venal motives of certain persons cause retardation of the entire country.

Without dwelling further on this issue, let me point out that the main foregone opportunity is that Russia, during the period from the late 1990s through the early 2000s, has enjoyed the realistic opportunity to reduce the income tax burden imposed on legal entities up to 10-15%; to forego all other taxes; to form the basis for its budget through imposition of leasing fees, excise taxes on alcohol, tobacco and luxury; and to apply protective duties while not being associated with the World Trade Organization ("WTO"). The second accusation is the collapse of scientific and technical complex and tolerance of the "brain drain." The third accusation is the imitation of educational reform. Finally, the fourth accusation is the resistance to evolutionary processes in social life.

Today, everything that could have still developed is smothered by taxes. Investment climate has become the same as in Antarctica. On the other hand, a daily depreciating piggy bank worth 0.5 trillion dollars has been sent to aid the development of the United States. However, let me come back to the question of time.

Understanding and formulation of the current moment are of significant importance. As rightly pointed out by S. P. Nikanorov, the inability to do so leads to "dominance of the situational, rather than historical perspective on the current events."[23] As he has aptly put it, this reminds us of the kids in a sandbox, "Oh, you've ruined my house! Then I'll break your shovel!"

Let's admit the Communist Party of the Soviet Union did this with the greater integrity than what we may observe in the annual presidential addresses. I refer my readers to the referenced source as an example of vision of the current moment.[24]

Description of the current moment is quite clearly regulated. This means:

- Analysis of the outcome of the goals achieved, as set during the previous period;

- Identification of the causes and errors which have not allowed to achieve the necessary results;

- Measures taken, their timeliness and effectiveness;

- Rationale for necessary adjustment in goals and their reinforcement through additional resources;

- Analysis of the current moment in light of the results achieved, changed circumstances, external influences, new understanding

---

[23]   S. P. Nikanorov, Experience of Historic Qualification of the Current Moment (Moscow, Russia: Concept, 2009), no. 24: 76. http://spnikanorov.ru/uploads/books/opyt.pdf.

[24]   S. P. Nikanorov, quoted in ibid, 76-90.

of the problems, as well as scientific and technological achievements;

- Rationale for new goals and modification of goals that existed previously;

- Tasks and problems arising therefrom;

- Justification of the available funds for the purposes of achieving the goals that have been set;

- Determination of persons and entities responsible for implementation of the goals that have been set;

- Establishment of deadlines and order of reporting for the responsible persons and entities;

- Presentment of a model of the expected future upon implementation of the goals that have been set.

. . . .

Chapter 4. Organizational fundamentals of economy.

. . . .

§9. The hypothesis of existence of two different types of crises.

At this time, this paragraph shall be presented as the hypothesis requiring further verification. Verification will require further research which I hope to implement in the future. I have determined that the cyclicality of the economic system in time should equal 20-25 years. First and foremost, this is due to generational changes and, secondly, due to changes in the technical generations in production. That is, the cyclicality of the system is associated with the revolutionary changes occurring with its main elements. If the cycles of changes in the human generations are intertwined, this should be somehow reflected in the industrial and social revolutions, as well as the crises. However, to test the validity of this assumption, one ought to clearly understand one important thing: not all crises are crises, that is, not all crises having economic undertones are economic crises. The fact of the matter is that the majority of the crises today is not dictated by internal failures within the economic system but rather introduced into the economy by the external actions of the civilized nature. Features of such civilization crises are expressed through ignoring the economic laws and processes as follows:

1) Price formation ceases to be a wave process around the cost, it takes a form of the "inflated bubbles" under the influence of

psychological processes, advertising and artificial excitement, which sooner or later burst (and most often, at the very moment wanted by someone).

2) Finance and credit policy ceases to be stable whereas:

    a— it ignores the establishment of the scale of prices for gold;

    b— living in debt becomes the norm, so that debts become uncollectible;

    c— the growth rate of the financial sector's profitability not only exceeds the profits earned in production, but also becomes an obstacle for the industrial development;

    d— the tax system becomes the robbery sanctified by the state.

3) Regular fraudulent activities by large companies and financial corporations consisting of manipulation with financial statements, concealment of an amount (volume) of devalued securities, misinformation of the economic community regarding available assets. Exposure of these scams is always fraught with the crises.

4) Spur-of-the-moment political ambitions of certain leaders seeking to fulfill their unrealistic political campaign promises or to "inflate" social programs for their next election term. This may also include incompetence of these leaders and their desire to participate in small victorious wars.

Civilization crises have no relation whatsoever to the economic cycles; therefore, their rate of recurrence is random and spontaneous. Economy here has no determinative character but rather has the subordinate one. Civilization reasons should be viewed as an external factor, the impact of which leads to the crises that look like economic in nature. Understanding of true causes of the crises inevitably leads to one's deeper understanding of the regularity of the processes, thus to their anticipation and methods to overcome the same.

I am able to partially verify my hypothesis regarding the 20-25 year long socio-economic cycle as follows. If one focuses on Russia and takes 1914 as the starting point, one will see that, by this time, the internal combustion engine begins to massively displace the steam engine. Aviation ceases to be exotic and becomes the real component of the military and transportation service. Automobiles begin to displace the animal-drawn cartage. There appear automatic weapons and tanks. They start using electricity on the industrial scale.

Technical rearmament disturbs the nearly centennial peaceful era in Europe. Russia experiences World War I, two revolutions and Civil War over the period of the next 7 years.

It is 1939. The country has concluded the period of industrialization and collectivization of its agriculture. Science undertakes testing of the rocket engines. They raise the question of using the atomic energy for civil and military purposes. Electrical energy becomes the main energy resource of the industry. They have now mastered the assembly-line production. Military equipment has lived through its 25-year cycle of modernization, acquiring qualitatively different characteristics. The same may be said of plant equipment. Genetic and artificial selection research is being conducted in the fields of Biology and Agriculture.

Six years of World War II will dramatically change the country's social face. The Stalinism will continue to exist; however, its days are practically numbered. Repressions still take place, but they are no longer widespread and cause mixed reactions in the community.

It is 1964. Space era and era of the automated control systems has begun. Jet aviation is displacing the "aviation of propellers." Railway transportation switches to the electric propulsion. First atomic power plants are being built. There appear machine tools with the programmed control (Programmed Numerical Control). In the Soviet Union, 1964 is associated with the completion of the period of the "Khrushchev thaw," dismissal of N.S. Khrushchev from all state offices and entry into the long period of "stagnation," as well as the growing influence of the "neo-Stalinists."

Change in the country's leadership is marked by the beginning of the Kosygin's economic reforms, which were nullified by 1970. Discovery of oil fields in the Western Siberia traps the country as a hostage of the economic model of "resource curse."

It is 1989. This is practically the year when the Perestroika begins. The world is on the verge of the computer revolution. The world develops new electronics and robotics. Product quality is at the forefront. New methods of the production organization are being developed. The Soviet Union is still considered a superpower, but its technological weakness has increasingly made itself felt. The country has steadily become the "raw materials appendage" of the West.

During this period, the mankind visited the Moon a few times and created the space shuttles. A space station with the cosmonauts on board is constantly working in space. There now exists a personal computer, and a mobile phone has been invented.

The Perestroika was ended with the coup of 1991, collapse of the Soviet Union and execution of the Parliament in 1993. The country rejects socialism and restores capitalist relations throughout its territory. Elections of 1996, accompanied by pledge auctions, irrevocably create the "thieves' capitalism" in the country.

Over the last 25-year period, we have experienced two civilization crises in 1998 and 2008-2011, which we do not take into account. Noteworthy is the foreign component of all aforementioned crises. Germany has always been present therein as Russia's "twin"; in the first two cases as the military adversary. In 1961 the crisis is preceded by the conflict with the West over the West Berlin and construction of the Berlin Wall around it. 1989 is marked by destruction of the Berlin Wall and subsequent reunification of the two Germanys.

In varying degrees, the United States is present in all aforementioned turning points. The Caribbean crisis of 1962 and the conflict of 1961 around the West Berlin associated therewith largely contributed to the change in leadership in the Soviet Union.

The end of 1980s of the past century is characterized by the artificial drop in oil prices to $15 organized by the United States along with the Saudi Arabia, the Soviet Union being caught into a new span of the arms race, as well as the imitation of success in the SDI program (Strategic Defense Initiative) associated with the conversion of space into the military polygon. This cycle is not Russia's prerogative. Countries of the Eastern Europe, Japan and China get involved therein through Russia. Countries of the Central and Western Europe get involved therein through Germany. The United States' participation in this cycle makes the same global in nature. These cycles also demonstrate the importance of Russia in the global process regardless of the state of its economy in any given period. The events of 2014-2020 will demonstrate whether it is possible to extrapolate this into the future.

I am not sure the 25-year long period will be so strictly observed in the future. However, I have no doubt we will experience significant changes during the period from 2014 to 2020.

Information technologies, computers, cell phones, and GPS systems have dramatically changed the economic and social processes. Space satellites provide communications on Earth, conduct surveillance, as well as monitor climate processes and execution of agreements in the military. Human presence in space has become ongoing. Human DNA has been decoded. Nanotechnology has crossed over into reality from the research and development stage. The mankind has realized the danger of nuclear explosions on Earth and in space. Demands for humane military technologies have sharply increased.

In the next 25 years we should expect significant advances in robotics and creation of artificial intelligence. Medicine will conquer cancer and significantly advance in the possibilities of prolonging a human life. Eradication of diseases on a genetic level will become possible.

It is way harder to predict social and political changes. Clearly, we are the witnesses to the sunset of the Yeltsin-Putin era. A new generation of politicians will come into place, who will attempt to get rid of the dirt of fraudulent elections and lead the fight against corruption differently. There will be an end to the state monopoly in the media. Russia will actually begin initiating radical, progressive reforms.

Imagine that we are now in the late 1988-early 1989. Trends are obvious enough. The arms race has reached its peak. Both the U.S. economy and the economy of the USSR are overwhelmed by military expenditures. Nuclear holdings threaten the lives of the entire mankind. The world is tired of the military threats and demands disarmament.

The socialist camp headed by the Soviet Union is in crisis. It is no longer possible to hide the stagnant trends. The announced Perestroika has no objectives and has not led to any results. A. D. Sakharov made a statement regarding the need for revision of the treaty of alliance on formation of the USSR.

Based on these facts, many events of the 1989-1996 period were realistically expected and anticipated, but only by way of integral characteristics. No one would be so forward as to announce the specific events of destruction of the Berlin Wall or the coup of 1991 and, in my opinion, it is impossible to do so.

If one goes back to 2014, then on the international scale, one discovers the existence of debt crises in the United States and Western Europe, which have reached their critical indicators. One discovers that the U.S. military actions are underway in Iraq and Afghanistan. Syria has become the suppurating mess. It is highly likely that the United States will take initiative in resolving the crisis in this country as well.

Transformation of the United States into the oil exporter and lifting the embargo on Iran will allegedly hurl the oil market, which will affect the entire global economy. Artificial price control on gold will result in gold prices skyrocketing up to $3000 or even up to $5000 per ounce. By then, financial geniuses will manage to turn their papers into hard currency.

Conflict between the government and the opposition in the Ukraine will not resolve positively on its own. Anything may happen there, a civil war, intervention of international forces, and partition of the country into pieces.

I envision that, this time, the United States will become the source of the global crisis. However, it will change Russia radically. It is logical to expect the United States will announce its default on the galloping debt of the country. Other crisis phenomena in the politics and economy of the country may contribute thereto. Separatist movements in the United States on part of some states are much stronger than expected by many living abroad. A crisis may very well exacerbate these moods. Default in the United States will be followed by a chain of defaults in the countries of the Western Europe. Further, I do not rule out that forced terrorist attacks may be organized on their territories with the goal of redirecting the discontent of the masses away from the economic and historic problems. In a situation of a large-scale global crisis one cannot extrapolate existing problems and their solutions under normal conditions. One may compare such periods to a plane crash when the system is able to withstand one but not more than two failures in order to function. However, when there are more failures and they overlap one another, disruption in the system's functions grows exponentially. What the system has coped with for many years no longer works, and disaster becomes inevitable.

I will not keep to myself that I am aware of the predictions of certain well-known psychics that the 44th president of the United States will be the last president of this country, and that the country will either fall apart or start losing its status as the world leader during his administration. Let these predictions be on those psychics' conscience. From today's perspective, such development of events seems highly unlikely. However, because I make an assumption that this will be one of the most serious global crises ever known to the mankind, this historical approach should not be ruled out.

A. P. Nazaretyan[25] rightly observes, "Since our secondary school years, we have been preached the conventional theory that the revolutions are implemented by the impoverished masses. Anecdotes were narrated in our history lessons, resembling the one about the Queen of France Marie Antoinette, who upon learning the day before the rebellion started that peasants had no bread, allegedly stated "Let them eat pastries" . . . And yet, in the last century a distinguished historian A. de Tocqueville convincingly proved that the revolutions always follow the years of economic growth and enlargement of political freedoms and opportunities. In particular, the standard of living of the French peasants and artisans was the highest in Europe during the 1780s of the 17th century, on the verge of revolution."

---

[25] A. P. Nazaretyan, Aggression, Morality and Crises in the Development of the World Culture. Synergy of the Historic Progress. Series of Lectures (Moscow, Russia: Heritage, 1996), 6-7.

Subsequently, de Tocqueville's conception was slightly clarified, ". . . this period is followed by a short period of a relative downturn often associated with an unsuccessful war. A rapidly widening gap between the passively growing social expectations and the decreasing objective opportunities increases the tensions fraught with explosion."[26]

When defaults are announced, Russia will be left without its illusory piggy bank worth half a trillion dollars, and Central Bank will lose a significant part of its reserves. Along with the drop in oil prices, the Putin's regime will turn out fully bankrupt in the political, economic and social sense. This will be a well-deserved outcome to the regime established after 1993.

Today, no one is in the position to say how a change in power will occur. However, this path has no alternatives. A new government, whatever it may be, will need to solve the following problems to emerge from the crisis and pave the way toward the real development:

1. Change the country's political system through achieving the actual separation of powers and doing away with the latent form of monarchy;

2. Ensure the actual will of the people is carried out through open and fair elections;

3. Put an end to corruption in the highest echelons of power;

---

[26]   A. P. Nazaretyan, quoted in ibid, 6-7.

4. Radically reduce taxes and redirect the tax system toward leasing payments;

5. Finance and credit system should take the right place stipulated by the laws of Economics, i.e. serve the production activities as well as scientific and technical progress.

I will not continue this list; otherwise, this book will be transformed into a certain political party's program. This list is sufficient to understand the complexity of the tasks ahead, their directivity and the power of resistance that will accompany them in the process of their implementation.

In conclusion, I would like to draw my readers' attention to one additional chronological momentum. World War I began for Russia on August 1, 1914. World War II began on September 1, 1939. N. S. Khrushchev was dismissed from his state offices by the Oktyabrsky Plenary session of the Central Committee of the Communist Party of the Soviet Union, which was held on October 13, 1964. It was preceded though by the Presidium of the Central Committee of the Communist Party of the Soviet Union, whose decisions formed the basis of the documents of the Plenary session. 1989 was abundant with events, the most significant of which were the First Congress of the People's Deputies, which was held from May 25 to June 9, 1989; and the Second Congress, which was held from December 12 to December 24, 1989.

However, the key point, in my opinion, was the fall of the Berlin Wall on November 9, 1989.

In light of the foregoing, I will be surprised as much as the readers of this book, if the anticipated events occur in the first or, more likely, in the second decade of December 2014. I will be surprised because I am not a psychic but merely trying to explore the economic and social processes. I cannot yet explain the reasons for such rigid cyclicality equal to 25 years and 1 month.

Bibliography

1. Aksenov, G. P., Vernadsky, V. I. About Nature of Time and Space (Moscow, Russia: Publishing House Librokom, 2012).

2. Balatsky, E. V. "Evolution of the Concept of Time in the Science of Economics," Federal Online Edition "Capital of the Country" (May 18, 2010). http://www.capital-rus/articles/article/177289

3. Bertalanffy, Ludwig von. General System Theory: Foundations, Development, Applications (New York: George Braziller, 1968).

4. Bertalanffy, Ludwig von. General System Theory—a Critical Review in Studies in General System Theory (Moscow, Russia: Publishing House Progress, 1969).

5. Geltser, Y. G. System and Technical Fundamentals of a Life Cycle of Engineering Design of Aquatic Structures (Moscow, Russia: SvR-ARGUS, 2006).

6. Geltser, Y. G. "Cycles and Their Asymmetry: Economic aspect," Journal Concepts, no. 2 (21) (July 16, 2008).

7. Grinin, L. E., Korotayev, A. V., Tsirel, S. V. Development Cycles of Modern World System (Moscow, Russia: Publishing House Librokom, 2011).

8. Guriev, S. M. Myths of Economy. Misconceptions and Stereotypes Distributed by the Media and Politicians (Moscow, Russia: Alpina Business Books, 2006).

9. Kobyakov, A. B., Khazin, M. L. Sunset of the Dollar Empire and the End of "Pax Americana" (Moscow, Russia: Veche, 2003).

10. Kuzyk, B. N., Ageyev, A. I., Dobrocheyev, O. V., Kuroyedov, B. V., Myasoyedov, B. A. Russia in Space and Time (History of the Future) (Moscow, Russia: Institute of Economic Strategies, 2004).

11. Meadows, D. Thinking in Systems: a Primer (Moscow, Russia: BINOM. Knowledge Laboratory, 2011).

12. Meadows, Donella H. Thinking in Systems: A Primer (White River Junction, VT: Chelsea Green Publishing, 2008).

13. Molchanov, Y. B. Four Concepts of Time in Philosophy and Physics (Moscow, Russia: Publishing House Science, 1977).

14. Nazaretyan, A. P. Aggression, Morality and Crises in the Development of the World Culture. Synergy of the Historic Progress. Series of Lectures (Moscow, Russia: Heritage, 1996).

15. Nikanorov, S. P. Experience of Historic Qualification of the Current Moment (Moscow, Russia: Concept, 2009), no. 24. http://spnikanorov.ru/uploads/books/opyt.pdf.

16. Ostalskiy, A. V. A Brief History of Money (St. Petersburg, Russia: Commercial Publishing House Amphora, 2010).

17. Taleb, N. N. Black Swan. Under the Sign of Unpredictability (Moscow, Russia: Calibri, Azbuka—Atikus, 2013).

18. Masanao Toda, Emir H. Shuford, Jr. "Logic of Systems: Introduction to a Formal Theory of Structure," General Systems: Yearbook of the Society for the Advancement of General Systems Theory, vol. 10 (1965).

19. Vernadsky, V. I. Problems of Biogeochemistry (Moscow, Russia: Publishing House Science, 1980).

www.ingramcontent.com/pod-product-compliance
Lightning Source LLC
Chambersburg PA
CBHW070714180526
45167CB00004B/1472